What are the threats to Earth from Space?

Table of Contents

Chapter 1. - We know that the Big Bang of 13 billion years ago saw the birth of the sun and many other planets. The Big Bang and historical evidence of the sun from Newgrange to Maeshowe. What powers the Sun?

Chapter 2. - The sun which is 90 million miles away from the earth is our own star. The sun's corona, eclipse, the sun's core and makeup. Solar weather. Sunspots, solar maximum and solar minimum.

Chapter 3. - On 15th February 2013, people on their way to work in Chelyabinsk, Russia were to witness an event that happens every fifty to one hundred years. Asteroids, meteorites Chelyabinsk. Chebarkul meteorite. The Yarkovsky Effect. The Tunguska Event, eyewitness testimony. The Yucatan Peninsula meteorite or comet? Comets and their components.

Chapter 1

We know that the Big Bang of 13 billion years ago saw the birth of the sun and many other planets. It just took an instant for the birth of the universe and the universe has been growing or expanding at the speed of light ever since. There are 100 billion galaxies within the universe and our galaxy is just one of them with over 100 billion stars. The sun is also just one star in the universe with Mercury and Venus being the closest planets to the sun. The sun is so hot that the planets Mercury and Venus have scorched surfaces due to the intense heat.

The big bang (Image: Mehau Kulyk/SPL/Getty Images)

All life on earth owes its existence to the sun. Without the sun the earth would be a barren planet made of rocks and with no life. We know from history that the ancient people worshipped the sun as they knew that the sun's heat and light was imperative to their existence. Newgrange in Ireland and Maeshowe on Mainland, Orkney, Scotland are testaments to the ancient's beliefs. These monuments can be called calendars and observatories of the sun. In Orkney the summers are hot and full of light while the winter is cold, dark and wet.

The weather and temperature is very similar to Newgrange. 5000 years ago a civilisation lived and thrived on Orkney and the ruins of Maeshowe were built by the people living on Orkney. Maeshowe was built 1000 years before the Pyramids of Egypt were constructed and Maeshowe and Newgrange were built by Stone Age architects. When Maeshowe was excavated the main chambers clay floor was covered in pieces of broken bones and skulls. We know from its remains that it was a place for burial and the dead. Like Newgrange a phenomenon occurs on the 21st of December at sunset of the winter solstice, the shortest day of the year.

Newgrange County Meath

The light from the sun shines into Maeshowe right through the entrance tunnel and illuminates the interior of the structure. Because this is the shortest day of the year and from this day onwards days are longer and light will increase. We can believe that the dead are awakened on this day. The calculation is incredible and tells us that in the Stone Age, the sun was acknowledged as being of vital importance to people's lives.

Maeshowe, Orkney, Scotland

Newgrange, located in County Meath, Ireland is a prehistoric monument built around 3200bc during the Neolithic Period (new Stone Age Period 10200BC-2000BC). It is older than the Pyramids of Egypt and Stonehenge. There is no conclusive proof as to the function of Newgrange but like Maeshowe in Orkney, Scotland it is aligned with the rising sun on the 21st December, the winter solstice with its inner chamber coming alight through the sun's rays. Newgrange stands on a circular mound 76 meters across and 12 meters high with inner earth and stone passageways and 3 small chambers.

Newgrange at Winter Solstice

Maeshowe at Winter Solstice

Computers today allow us to look back to see what our ancestors saw of the sun, moon and stars. It is believed that Babylonians were the first astronomers but when we look to Ireland in the Neolithic Period we can assume that there was a careful analysis of the sun and stars carried out. The people from this era observed a correlation between the sky and the earth. Stonehenge was built to track the cycles of the sun and moon. We know that Newgrange is older than Stonehenge and through analysis of Newgrange and its surroundings we can see a much bigger picture between the relationship of the sky and earth.

The River Boyne has mystical and historic significance for Ireland. The name the River Boyne means the river of the White Cow. In the time when Newgrange was built the Milky Way was known as the way of the White Cow. We can suggest that the River Boyne was seen as the river of the sky. Newgrange stands on the banks of the River Boyne and its name in Irish, Bru Na Boyne means the womb of the White Cow. On the shortest day of the year, December 21st, the ancient builders perhaps wanted the light to overcome the darkness. Or perhaps it is similar to Maeshowe for the bones of the dead to be re enlivened by the light of the sun.

White Cow Road

Maeshowe interior

When we look at analysis of the sun we can see that it is far from being consistent and it is constantly changing. The sun has a life of its own. To see the sun as it is, modern solar observatories magnify and filter the sun's light to give us a clear view of the sun's exterior. The sun is turbulent and boiling internally with a temperature of 6000 degrees centigrade on its surface. The sun is a vastly huge star as the earth could fit one million times into the sun.

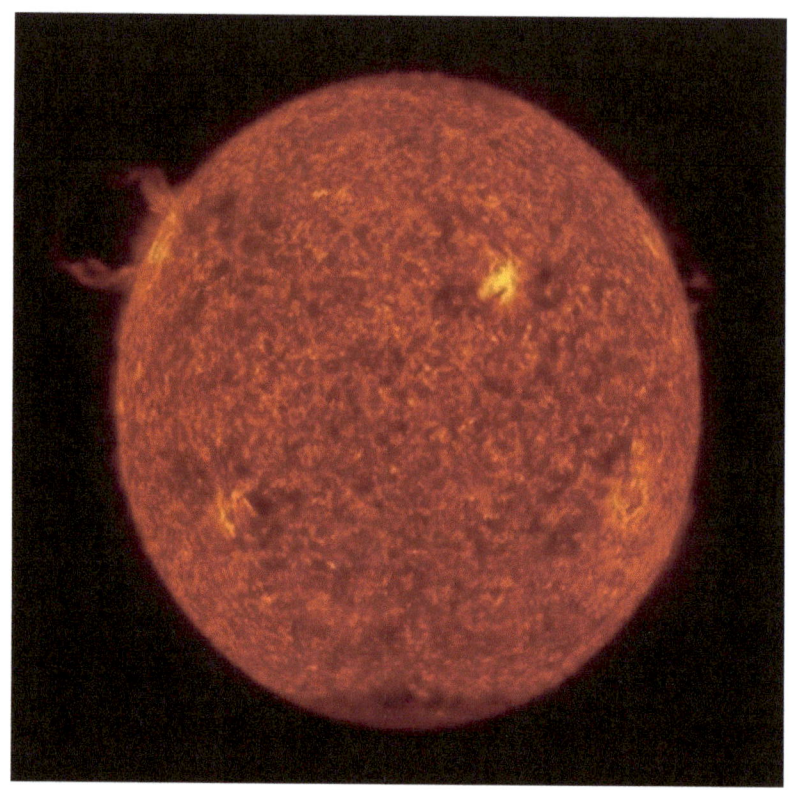

The sun's surface NASA's Solar Dynamic Laboratory 22 April 2010.

The sun is turbulent and boiling and never the same from one second to the next. The sun's surface bubbles just like boiling food under constant heat. What we would call bubbles or sun spots are at least 1000 miles across. The heat and light from inside the sun raises the temperature on the surface to 6000 degrees centigrade which would allow the sun to vaporise solid rock. Huge explosions happen on the sun that releases the energy of one million atomic bombs in seconds. The power of the sun comes from within the star as energy rises from within, travelling to the sun's surface.

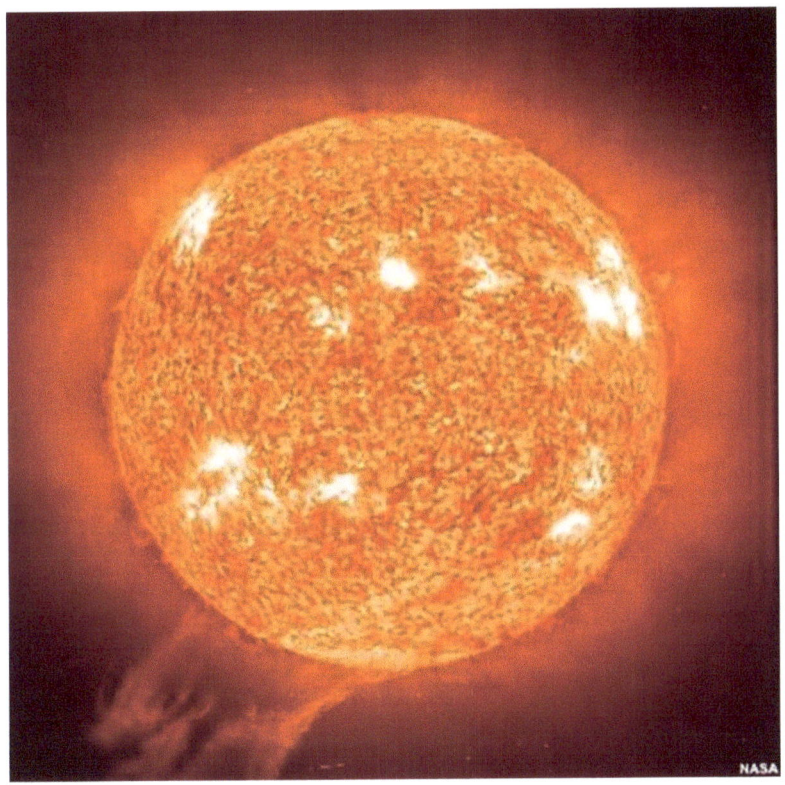

The Sun

If we could harness the sun's energy or power output for one second it would supply the world's energy demands for the next one million years.

What powers the Sun?

It was believed years ago in the times of the Stone Age that it could be coal that powers the sun but if that were the case the sun would burn itself out in a few thousand years. In the past people believed that the earth was a relatively new planet and was only a few thousand years old but when science emerged, through the analysis of rocks, geologists worked out that the earth dates back millions of years. It was understood by scientists that the earth was over one billion years old and through this understanding it was impossible for the sun to be made of coal as it would have burned itself out millions of years ago.

Through analysis of the light from the sun and passing the light through a prism, scientists found the chemical elements that make up the sun's light. What was found was the complete chemical formula of the sun. What the scientists learned was that the sun is composed of an enormous amount of Hydrogen with other elements such as Helium, Oxygen, Carbon, Nitrogen, Silicon, Magnesium, Neon, Iron and Sulphur.

We know that hydrogen is the most common element in the universe. The gas, hydrogen, can be seen floating in space, with hydrogen being the most common element for the creation of a star. The floating mass of hydrogen is called a Nebula. Within the hydrogen clouds floating in space which can be hundreds of light years across we can see stars being formed. Hydrogen is used in the formation of stars which it compressed as the gravity of the particles sucks in the hydrogen with the temperature of the hydrogen rising which is known as Nuclear Fusion.

Omega Nebula NASA

As these stars get bigger they start to spin and throw out debris that eventually forms a solar system. We see the earth and other planets spin in the same direction around the sun. The young developing star will eventually light up and become a new star. Every atom in everything around us was made in the heart of a star with the gas hydrogen being the starting ingredient.

Within the sun, hydrogen which is positively charged is forced to bond together to make helium due to the high temperature and pressure. The centre of the sun which is known as the core is boiling at 15 million degrees. The protons (sub atomic particles) are forced together so hard that they fuse (to reduce to a liquid or plastic state by heat). Energy is equal to mass times the speed of light by the speed of light ($E=mc^2$). The speed of light is a very big number. A small amount of mass creates a large amount of energy. That is the energy that powers the sun. Every second 5 million tons of the sun is converted into pure energy. The sun has been boiling for 5 billion years and it is only half way through its supply of hydrogen. The light produced at the core of the sun must travel half a million kilometres to reach the surface. It can take over 200 thousand years for the light to get to the surface of the sun and takes only 8 minutes to reach the earth. This is Nuclear Fusion. The H Bomb was man's first attempt in creating the power of the sun on the earth. It is hydrogen squeezed until it releases its energy. If we could control nuclear fusion it would solve the earth's energy problems. This is something scientists have been trying to do on earth which in effect is to create our own star.

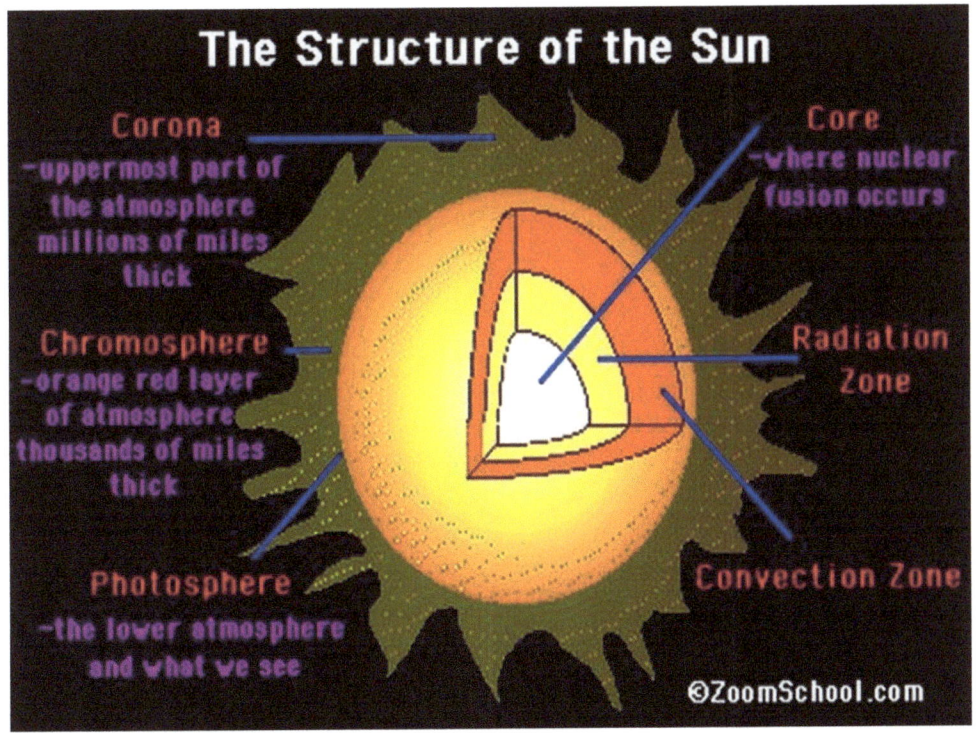

The Structure of the Sun

A major question for science is can we harness the suns power? If we could we would have electricity to power the earth for future generations. Through the work of the scientific community this may become a possibility in the future.

Within the sun there are other forces at work other than just light. The surface of the sun is torn apart by forces which cause sun spots. They appear as dark spots on the sun but these dark spots are normally the same size as the earth. A sun spot is an area of the sun that is extremely bright and not static. The sun spots look alive and we know from Galileo that the sun spots are all moving in the same direction. The sun is rotating and is turning faster at the equator than at the poles. From this information we get a further understanding of how the sun works. Records have been kept on sun spots over time. At times there are no sun spots visible on the sun. They come and go with an eleven year cycle. It is believed that sun spots do have an effect on the earth. We know from history that sun spots do affect the climate of the earth.

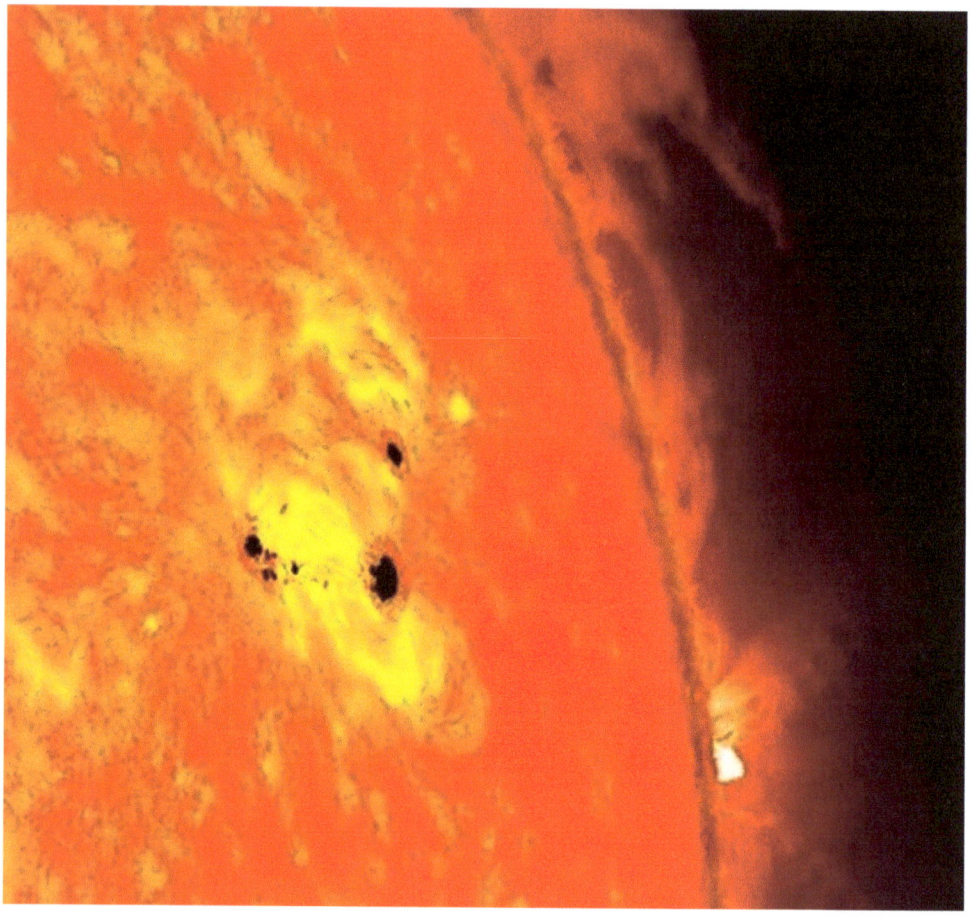

Sunspots. Image taken Feb. 2013 NASA Goddard Space Flight Center

In the 19th century the astronomer Robert Wanda made the discovery that by the end of the 17th century and the beginning of the 18th century sun spots had disappeared. There was no evidence of sun spots and this period had a prolonged cold stretch throughout the Northern Hemisphere which became known as the Little Ice Age. The Thames River in London, England froze over and civilisations disappeared. The earth cooled down. We can perhaps tell from this period that ultra violet light from the sun was absent and the absent heat from the ultra violet rays led to the cooling of the earth.

1895. February, Thames River frozen at Gravesend

When space craft get close enough to the sun and are able to record the sun's surface we can see sun spots glow a brilliant white. Gas spews from the sun spots and can be seen through x-ray frequencies. No one had seen the sun like this before. The astronauts on Sky Lab discovered this tremendous power of the sun- huge plumes of super heated gas at 1 million degrees centigrade. We can now see the explosive power of the sun, the solar flares and Coronal Mass Ejections which erupt from sun spots. The flares and C.M.E.'s are at tens of millions degrees.

Ultraviolet view of the sun NASA/Goddard/SDO AIA Team

Sun spots are the visual effect of the magnetic fields energy so strong that they stop the heat and light rising from the suns interior. Magnetic loops can be seen rising from the surface of the sun. The fields are stretched like elastic bands and cause the loops to snap and billions of tons of plasma are thrown out into space. Sometimes these flares or C.M.E.'s can reach the earth. They can hit the earth's magnetic field which throws them back into space but the impact on the earth's magnetic field can affect the planet. This is known as Space Weather. From this intermingling of energy from the sun and the protection from the earth's magnetic field we get what is known as the Aurora Lights, smashing through the magnetic fields of the north and south poles, (The Northern Lights).

Astronomers launched a NASA-funded rocket into the aurora borealis, which are colloquially known as the northern lights. The 46-foot rocket, the Magnetosphere-Ionosphere Coupling in the Alfvén resonator (MICA), launched from the Poker Flat Research Range, 30 miles north of Fairbanks, Alaska

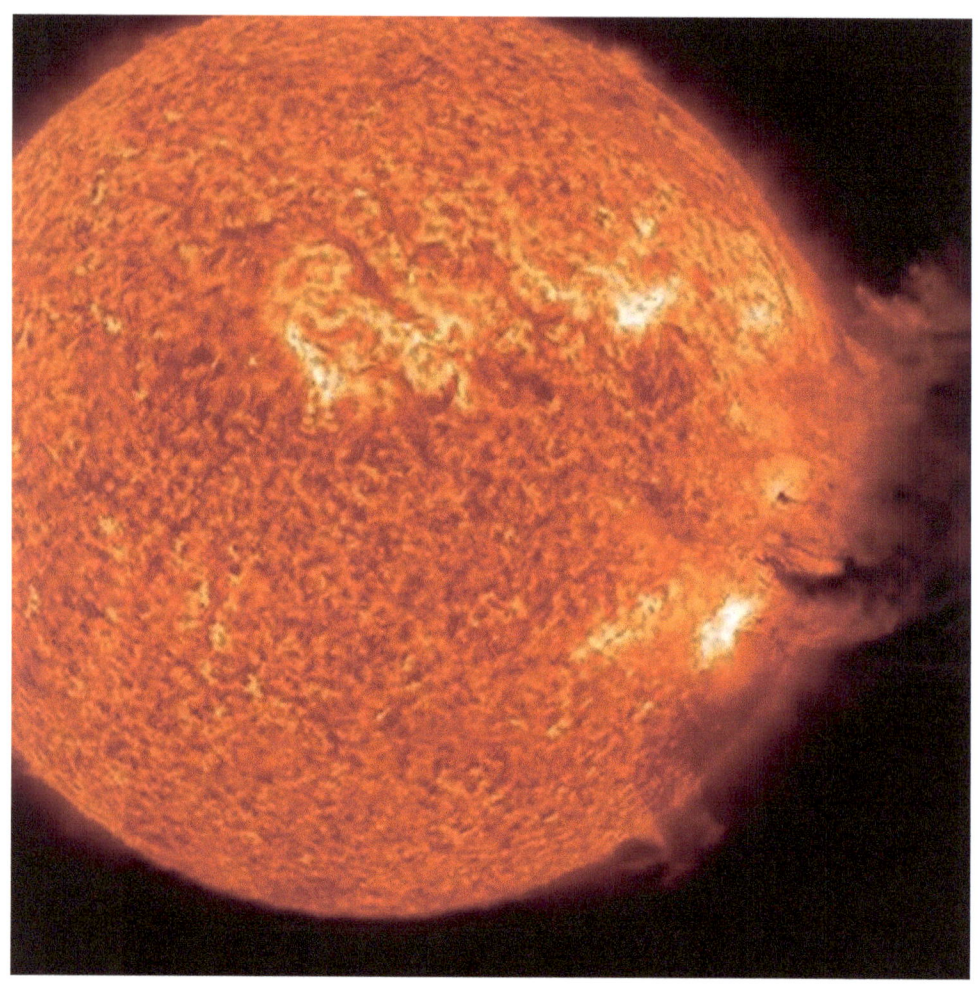

Coronal Mass Ejection. As viewed by the Solar Dynamics Observatory on June 7, 2011.NASA/SDO

Migrating birds can get lost as they lose their navigational skills and whales have been beached because of this, but the effect on the earth's electronics can cause very serious problems. The strongest storms can destroy satellites on which so much of the earth depends. Mobile phones to televisions, weapons guided systems; airplane navigation can all be disturbed by space weather as they rely on satellite communication. We still don't know why the disappearance of sun spots can cause an ice age.

There are sterling dishes (for harnessing the suns energy) set up in the States to receive energy directly from the sun produced by Sterling Energy Systems (S.E.S.). The dishes track the movement of the sun and produce energy.

Stirling Engine Systems

The sun has been with the earth for some time but it will also die in roughly 5 billion years. The sun will run out of hydrogen and when it does complete its hydrogen supply the sun will destroy the earth. As when all stars die, the sun will become a giant star as its core will shrink generating so much energy that the sun will balloon into the solar system destroying all of the surrounding planets. It will burn with tremendous heat and will perhaps swallow the earth. The sun will burn 2000 times hotter that it does now. It will stop burning and blow its remaining gas into space. The sun will end as will the earth.

The Sun

The sun, ninety three million miles away is coming so much closer to the earth. The light and heat that come from the sun are not as serene as we might think. Violent energy from the sun has affected the earth with its violent example in 1989. Quebec in east central Canada was hit with a violent solar storm emitting currents which overloaded the power station and left Quebec and most of Montreal without power. It took 93 seconds for the solar storm to hit and shut down power for over nine hours.

Quebec hit by solar storm March 1989

There are several active regions on the sun with sun spots causing the greatest and biggest space storms that have and can occur. Science has shown that sun spots are an explosion on the suns surface, exploding like thunderstorms and if the energy from a sun spot is directed towards the earth, it can cause a great deal of damage. Chaos and violence can be a definition for sun spots. It transmits a force so powerful, fierce and damaging that they can destroy billions of pounds of technology in seconds. Sun spots can also trigger solar storms.

The energy from a sun spot is a force of magnesium created within the sun. How we can comprehend this space weather is to understand the turbulent flow of plasma within the sun. It is like propellers spinning in different directions with energy flowing out from the sphere. There are enormously high temperatures in this process and the greater the revolutions within the sun the greater the emissions.

TRACE image of the sun (September 2000)

Due to the speed of the dynamics within the sun, the turbulence levels rise massively. The magnetic field of the sun is caused by magnetic energy within the sun having to find a way to the surface from the interior. Due to the twisting of the magnetic field within the sun stress is produced which eventually leads to an eruption of energy on the suns surface. Sun spots can be created and grow to critical levels due to the twisting of the magnetic fields. This same process can also create Coronal Mass Ejections (CME's) or Solar Flare. When a (CME) erupts billions of tons of the suns energy are hurled into space. CME's can be extremely dangerous if it is directed towards the earth. We have systems on the earth and in space that observes weather storms in real time as some of the events that occur on the sun can cause a direct challenge to the earth and the planets safety. A CME can knock out power on the earth and if the power on the earth goes out satellite system will also be gone.

The Advanced Composition Explorer Satellite (ACE) floating 1.1 million miles from the earth satellite has been protecting the earth since 1997. Once a solar storm emits from the sun and hits the Ace satellite we know here on earth that the solar storm will hit the earth approximately one hour later. Any solar storm that hits the ACE satellite will definitely hit the earth. CME's can also miss the earth and travel through space. The largest and most powerful CME is the X10 CME which can hit the earth at a speed of 2000klm's per second. An X10 CME or Solar Flare is the greatest and most destructive flare that can hit the earth. (X class is the biggest flare).

Advanced Composition Explorer Satellite (ACE)

When the solar storm hits the earth at total intensity the direction of the storm is of great importance to the earth. If the storm hits the earth at a northwards direction it is going to be a big storm but if it hits the earth at a southwards direction it is going to be a massive storm. The earth's magnetic field naturally repels storms that have a northward polarity. When the storms direction is southwards it is allowed through an open gate in the earth's magnetic field.

Chapter 2

The sun which is 90 million miles away from the earth is our own star. The activity on the sun, with its almost constant eruptions of super heated plasma and vast waves of radiation can at times be directed towards the earth. Every 11 years the sun goes through a peak in activity called the solar maximum. This is the high point in the sun as it goes from a period of relative calm to intense activity and then back to a period of relative calm. The sun also has its own faint atmosphere which is called the Corona.

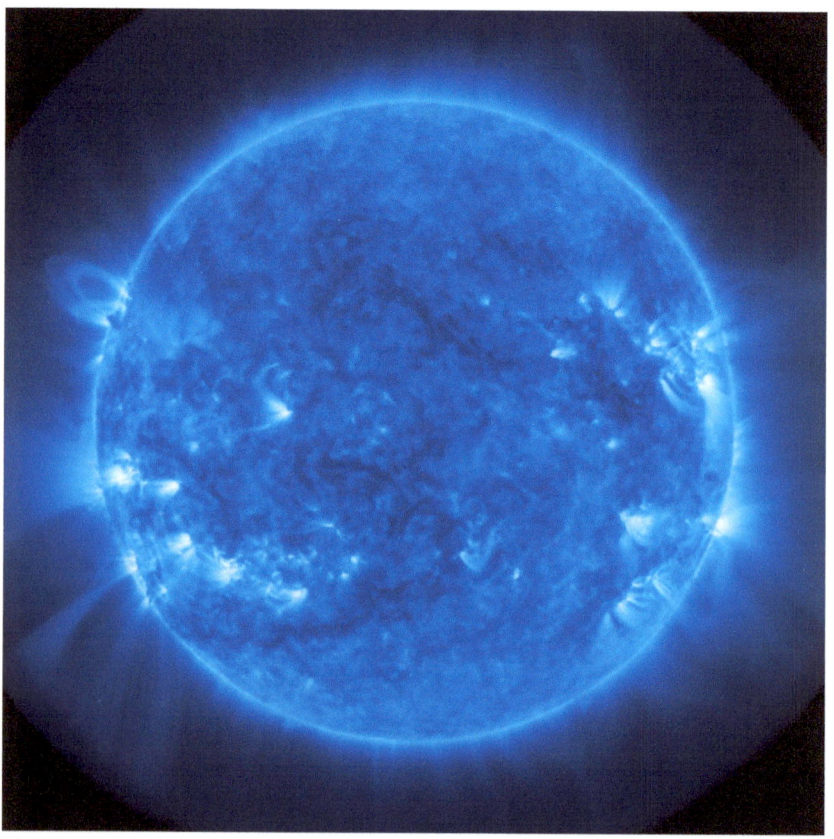

The Sun's Corona

The earth is the only planet in our solar system where you can witness a total eclipse of the sun. The reason for this is that the moon is 400 times smaller than the sun but it is also 400 times closer to the earth. When the moons orbit brings it between the earth and the sun it appears to be exactly the same size as the sun. The moon is able to block out the entire surface of the sun from our view. There is a total eclipse of the sun every 18 months.

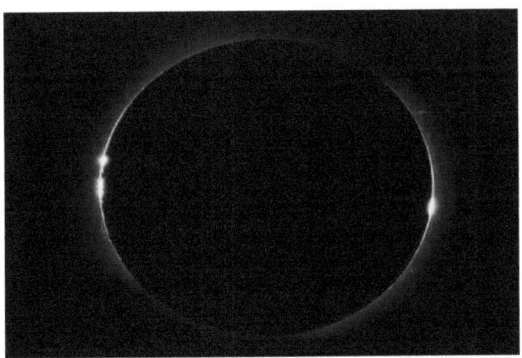

Solar Eclipse

The centre of the sun or the suns core burns as a massive 60 million degree furnace. The pressures and temperatures in the middle of the sun are so enormous that hydrogen atoms can fuse together. When this happens a small bit of mass in converted into a huge amount of energy. This sis called Fusion and this is the key to the life of the sun. If this process did not exist the sun would be a cold dead star and the earth would be a cold dead planet. Fusion is the key to life on earth.

The core of the sun is constantly generating endless energy and the particles of the suns energy are known as Photons. They are responsible for the birth of light energy from the sun. Photons or light travel through the sun and have to reach the suns surface. This is not an easy step as within the sun there is plasma between the core of the sun and the sun's surface. There is no direct route for the Photons to the suns surface so they are forced to navigate a route through the plasma. The photons move at thousands of miles per hour and there are hundreds of thousands of miles of plasma to cross between the suns core and the sun's surface. The Photons take a journey that should only last 2.5 seconds as they are travelling at the speed of light and there is nothing faster than the speed of light. It can take the Photon between 10,000 years to 1 million years to get from the core to the sun's surface.

When the Photon gets to the suns surface it takes roughly 8 minutes for the sun's light to cross 90 million miles before reaching the earth. This is the last 8 minutes of the sun's light.

Fusion in the sun's core never stops and is unending which ponders the question as to why does the sun's activity go up and down with the 11 year solar cycle? The reason for this lies in how the fusion reaction affects the sun's plasma. This in turn leads to the solar cycles. The heat generated by this reaction within the sun super heats the gas within the sun so the particles of gas are torn apart to form plasma. Just as the hot air on the earth around us rises the same reaction happens on the sun as the gases in the outer layers rises. This is called Convection. Gases get heated from below and rise up to the surface of the sun. Due to the fact that the gases, plasma is so hot it is also electrically charged. As it moves up and down with the convection currents it creates powerful magnetic fields.

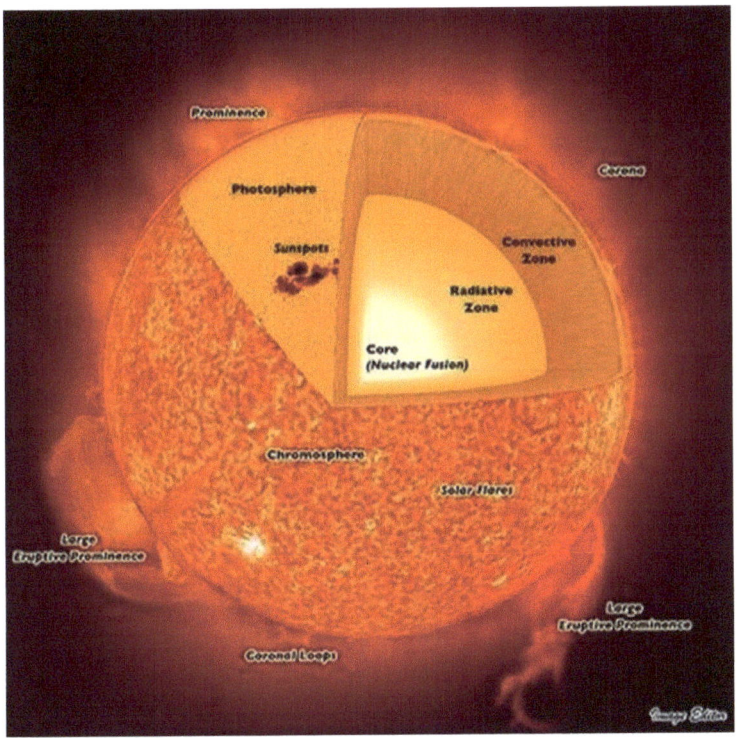

The core is where nuclear fusion takes place; this fusion causes enormous amounts of heat. This heat circles through the photosphere through convection currents.

The sun spins on its own axes just like the earth and this causes the plasma to also flow sideways. This in turn has a dramatic effect on the magnetic fields. The magnetic field lines get wound up and eventually it becomes so strong the magnetic field lines rises up and penetrate the surface of the sun. This is when we have a build up to a solar maximum. At times of solar maximum the magnetic loops break out from the surface of the sun drawing the sun's plasma with them. Some of the magnetic loops can be many times bigger than the earth. The sun does not stay in this condition. Eventually the magnetic fields disperse and re arrange themselves and goes back to a state of solar minimum where there is a nice ordering of the magnetic field. This happens every 11 years. The sun is 90 million miles away from the earth and the sun's cycle has great importance for us here on earth.

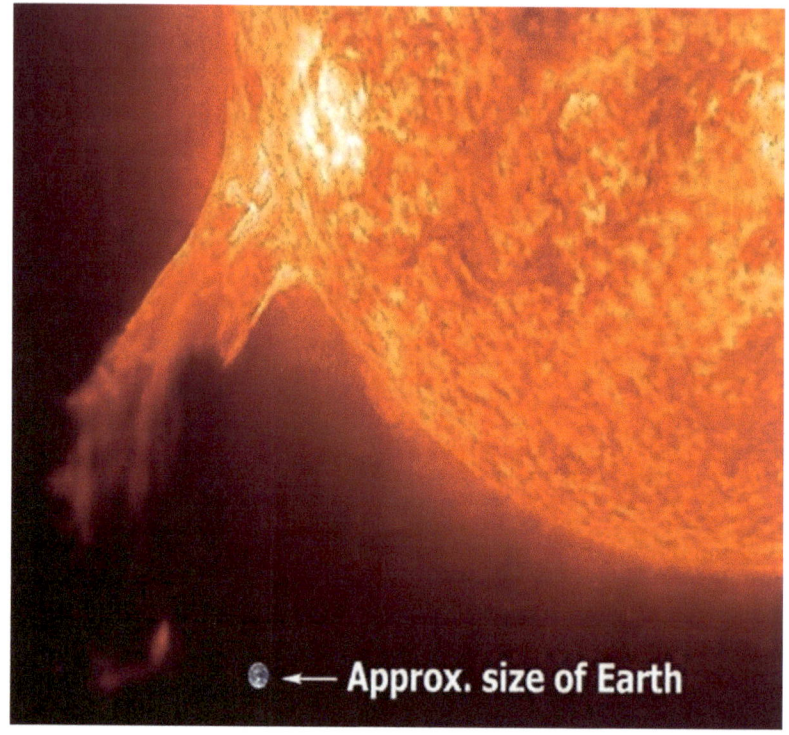

Solar Maximum SOHO EIT Consortium ESA NASA

At the current time when the sun is at its solar maximum with the sun expanding constantly into space the sun emits its solar winds. When the sun is at its solar minimum the wind is relatively quiet but at solar maximum we get vast streams of solar winds coming towards the earth. The solar wind is a constant stream of particles flowing out from the sun. The solar winds bombard the earth's atmosphere. Most of the sun's particles are deflected by the earth's magnetic field but some do get through to the earth and we can see the bright lights which we know as the Aurora Borealis or the Northern Lights. This is the solar wind made visible on earth. This is caused when the sun's wind encounters the earth's magnetic field. The effect causes the earth's atmosphere to light up in ghostly colours towards both poles. The best time to see the aurora is when the sun is at its solar maximum. The earth's magnetic field is thin and it does protect us from the solar storms.

Aurora Borealis. The Northern Lights. Image credit: unknown.

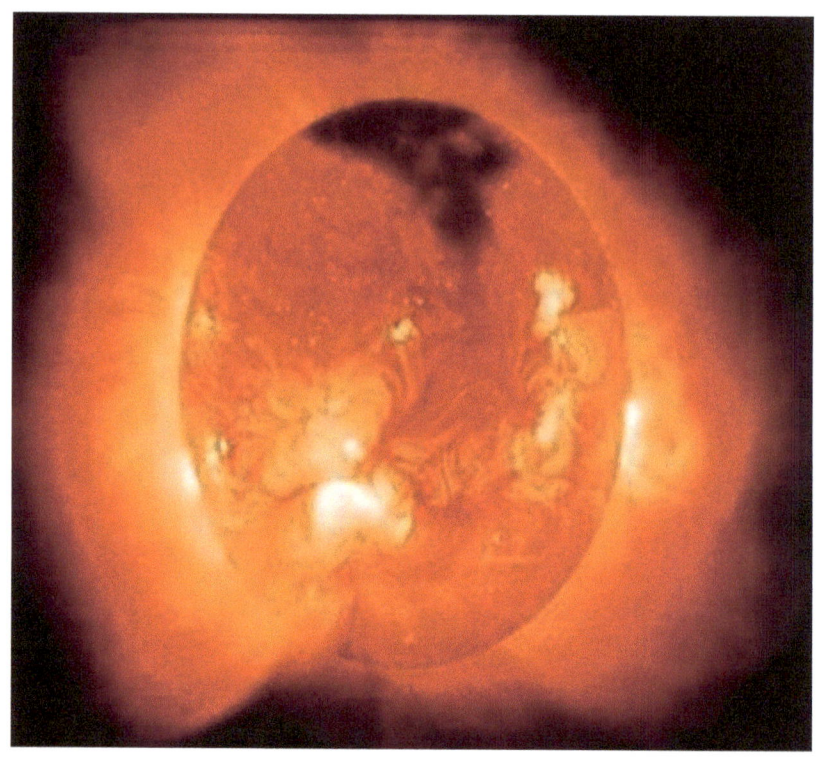

Solar winds set to pound the Earth in 2012. Image: NASA.

Solar Weather

Solar weather which is made up of radiation and particles can have alarming side effects for the earth. It was believed by solar scientists that we had a set pattern of cycles within the sun. Now there is a strong belief that the cycles of the sun are not set. On the surface of the sun we can see sun spots which were previously explored by scientists such as Galileo and other early explorers. We see sun spots as tiny dark spots on the sun. When Galileo mapped the sun spots he noticed that the spots were more numerous or that over time there were more sun spots or less sun spots then were previously recorded. It was also believed that the sun spots were tiny planets that were moving through space and what we were witnessing from the earth were these planets in front of our view of the sun. Through the use of telescopes solar scientists understood that what were thought to be tiny planets in front of the sun were in face part of the sun. New research suggests the sun spots are caused by the magnetic fields deep within the sun. The more active the sun the more sun spots appear.

Galileo Galilei

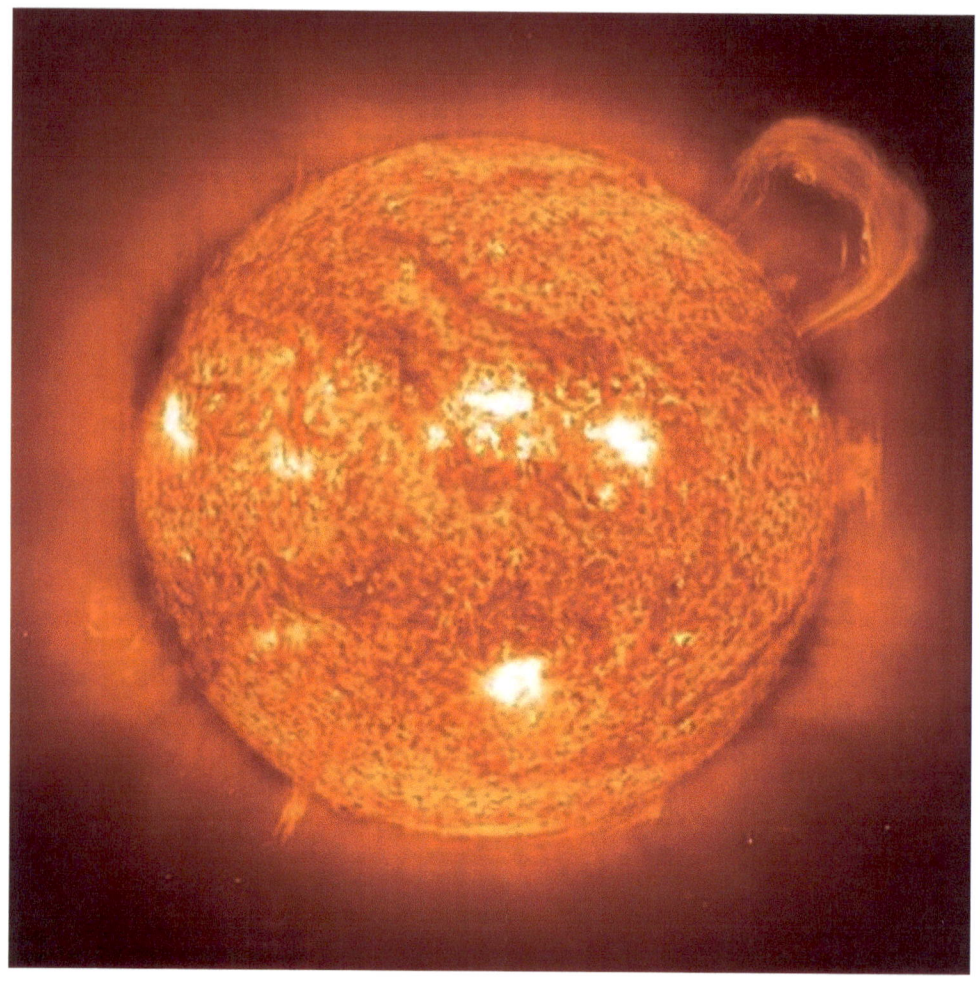

Sunspots. National Aeronautics and Space Administration

    The magnetic fields within the sun become entangled during the sun's solar maximum. The flows of plasma within the sun are disrupted which means that the hot material within the suns interior cannot rise to the surface. The results we see on the sun are the zones of cooler plasma.

    During the sun's solar minimum there might be zero to 5 sun spots but during the suns solar maximum this could rise to 100 sun spots. The magnetic strength of sun spots has been decreasing year by year over the past ten years. Solar scientists believe that in the not too distant future there might be no sun spots at all. This understanding has led solar scientists to believe that what was once understood of the sun having an 11 year cycle may be incorrect and that the sun's period of cycles might be much longer than 11 years. There are much bigger patterns in the sun's activity. The sun now might be heading for a grand minimum which is an extended quiet period of activity.

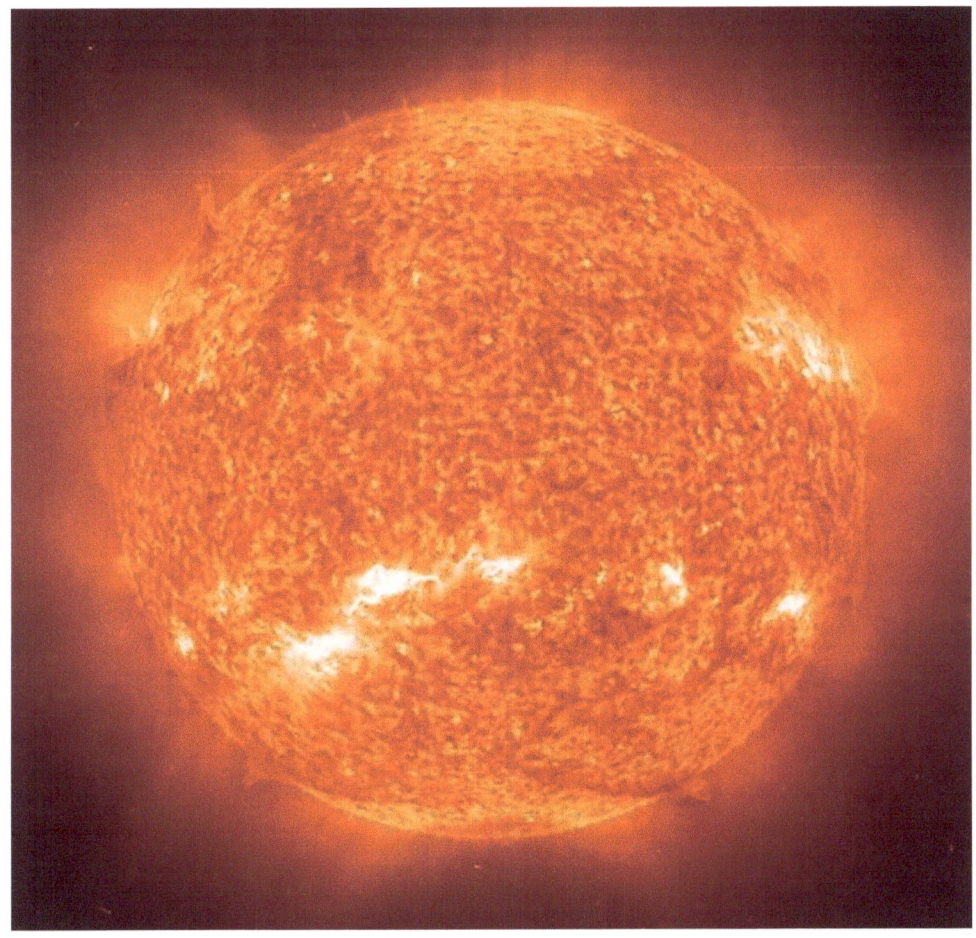

Sunspots taken by the SOHO spacecraft, 1998 (SOHO, NASA and ESA)

There has been as period recorded from the sun where the sun did go into a solar minimum where there were no sun spots at all. This happened

350 years ago and continued for over 70 years where sun spots almost vanished. Records show that the earth's previous temperatures in Europe decreased throughout the period when the sun spots almost disappeared. This period brought harsh winters to Europe and the Thames River in London froze solid. This period of time became known as the Little Ice Age. Through the decreasing number of sun spots we might be seeing evidence that there might be a bigger cycle in the sun's behaviour. Solar scientists now believe that the pattern of the sun which was thought to have an 11 year solar cycle might not the full story of the sun.

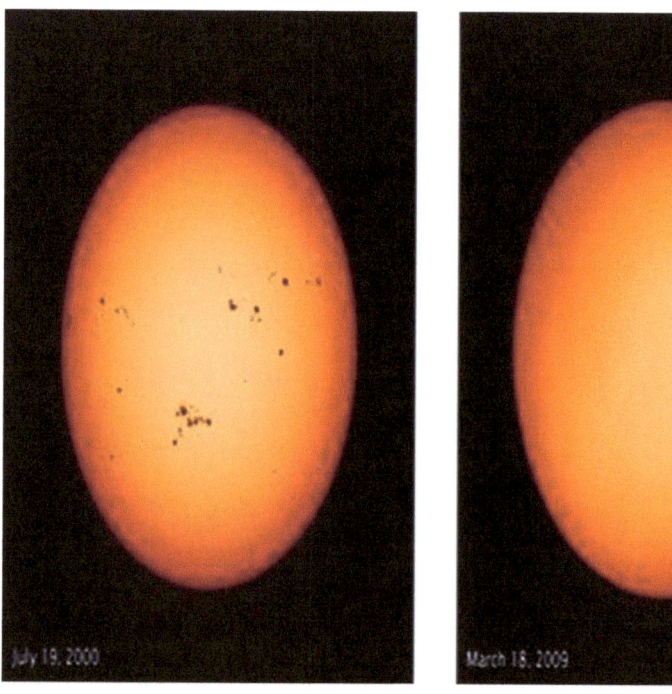

Solar Maximum and Solar Minimum. Solar and Heliospheric Observatory (SOHO)

Research on the sun has been greatly expanded due to the Rutherford Appleton Laboratory (RAC). RAC satellites are designed and tested before they are sent to space. Scientists are analysing the information these satellites beam down to earth. It is one of the most important centres for solar research in the world. We can't look directly at the sun without damaging our eyesight but a new fleet of satellites allow us to get a unique picture of the sun at RAL. Twin stereo spacecraft were launched by NASA in 2006 to observe the sun from two sides simultaneously. The solar dynamics observatory followed for years later. It is able to visualise the sun in high resolution for the first time. These new satellites allow for the sun to be analysed in much greater detail than was previously possible.

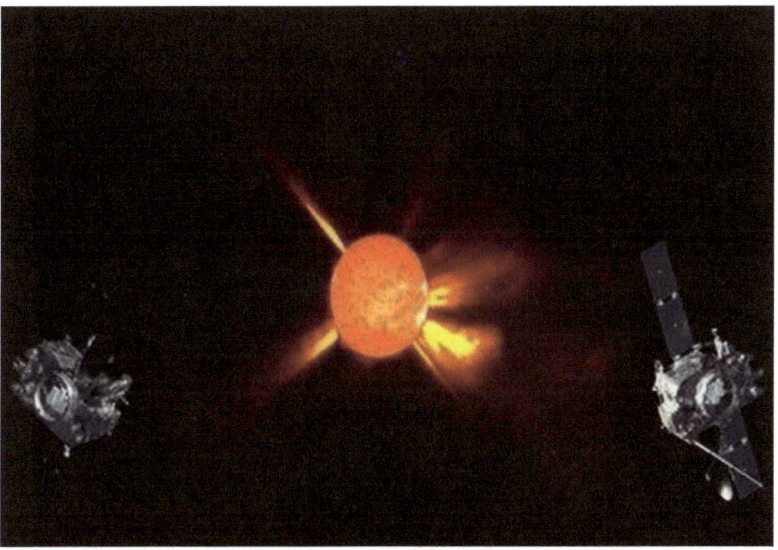

Twin Stereo Spacecraft. (Credit NASA)

Chapter 3

On 15th February 2013, people on their way to work in Chelyabinsk, Russia were to witness an event that happens every fifty to one hundred years. A meteorite, which looked like a fireball, brighter than any of the planets and brighter than the sun, was recorded on the cameras many have in their cars for insurance purposes. The meteorite exploded with the power of thirty Hiroshima's and surprised and terrified the residents of Chelyabinsk. A shockwave followed the explosion roughly one minute later blowing in the windows of over four thousand premises across the region. There were over 1500 injuries mostly due to broken windows, and it was the most powerful meteor strike for more than a century.

Meteor Showers Over Russian City of Chelyabinsk

What we learn from the Chelyabinsk meteorite is that the Earth is very exposed as we live in a vast, violent cosmos. There was no warning that a meteorite was heading on a collision course with the Earth. The meteorite was too small to detect and it caught the world of science by total surprise.

Many questions arise from this meteor strike. Meteors are normally tracked in space as they represent great danger to the Earth and for this reason they are closely monitored. Questions arise as to where this meteor came from within Space and is there a possibility that the Earth will be hit again. The greatest concern is can we protect our planet from further strikes.

Taken on a highway from Kostanai, Kazakhstan, to Chelyabinsk region, Russia

Scientists, locals and enthusiasts have descended on Chelyabinsk looking for pieces of the meteorite and it has been a hard task to find any fragments. The largest part of the meteorite weighing 570kg crashed into a lake in Chebarkul, and was recovered on 16th October 2013 by divers working on the lake.

Chebarkul meteorite recovered from lake

The large fragment created a six meter-wide hole in the ice. Smaller fragments have been found in the ice surrounding the hole. The meteorite has now been named as the Chebarkul meteorite after the city and lake it crashed into.

Approximate place of falling Chelyabinsk's meteorite in the Chebarkul Lake

No one in the scientific world saw the meteorite making its way through Space towards the Earth. Due to the size of the asteroid, a few meters across, it was undetectable. Scientists could not predict its impact on the Earth but it did excite scientists all over the world as its impact was recorded on many cameras.

From the recorded data we can see the meteorite break apart when it enters the Earth's atmosphere. As the meteorite breaks apart we can see that it releases some of its orbital energy as the Earth's atmosphere is denser making it more difficult for the meteorite to travel. The heat generated by the compression of air in front of the body (ram pressure) as it is travelling through the atmosphere is immense and most asteroids burn up or explode before they have a chance to reach the earth. For the people of Chelyabinsk this is a very positive feature for if the meteorite had not broken apart and had crashed to the Earth in its total size there would have been a devastating effect on the area.

Chelyabinsk Meteor 2013

As the meteorite breaks apart it releases a huge flare up followed by a shock wave that was detected over 15,000 kilometres away. It took 16.5 seconds from entering the Earth's atmosphere before breaking apart. The meteorite gave one of the largest infrasonic (low frequency sound) waves ever recorded. It was detected by monitoring stations from Alaska to Antarctica. The pressure wave (seismic wave) was seen all over the world. The colossal energy release was 500 thousand kilotons. The kiloton recording from the bomb dropped on Nagasaki in 1945 was 21 kilotons.

(Itar Tass) News Agency

The meteorite entered the Earth's atmosphere at roughly 17.5 km's per second and from the recordings we can combine the energy with the velocity and calculate the mass of the object. From the calculations we can tell that the meteorite was roughly 15 meters across. This was indeed a very rare event occurring on the earth and captured by many cameras.

Meteorites are formed in deep space. They are parts to much larger bodies called asteroids. Their sizes can vary from small parts of just a few meters to much larger sizes of up to 900 kilometres. Our solar system was created over 4.6 billion years ago from a nebula (from the Latin for cloud, an interstellar cloud of dust consisting of hydrogen, helium and other ionized gases) and asteroids are the leftovers of the material from which stars and planets were created. Millions of asteroids circle the sun and their trail is called the Asteroid Belt. Along the asteroid belt collisions can occur which throw an asteroid out of its set alignment and from the collisions small fragments are created. These fragments have a new path which allows some of the fragments to head towards the Earth. In the night sky we can often see the small pieces or fragments burning up in the atmosphere becoming meteors or shooting stars. When one of the larger pieces does not burn up in the atmosphere and reaches the Earth it is called a meteorite.

In this artist's concept, a narrow asteroid belt filled with rocks and dusty debris orbits a star similar to our own sun. Image credit: NASA/JPL-Caltech

A meteorite is a solid fragment from an Asteroid that originates in outer space, reaching the earth's surface and surviving the impact. It falls through our atmosphere and hits the ground to be recovered. Some of these rocks that have hit the earth have been dated older than the earth itself dating 4.6 billion years.

The meteorite that hit the ground in Russia, the Chebarkul meteorite is believed by scientists to be older than the earth. From analysing the components within a meteorite it is possible to tell the age of the components that make up the rock. The meteorite is made up of several components including Chondrites which are stony meteorites. They were free floating around the very young sun before there were planets and they slowly came together when there were small grains and dust in the early solar system combining to form asteroids. After this process they formed into larger objects until they eventually formed into planets.

Meteorite name: Chelyabinsk
Country: Chelyabinskaya oblast, Russia
Class: Stone, chondrite (LL5)
Date of fall: Fell February 15, 2013 09:22

Today the Earth is growing by over 40,000 tons per year and from this we know that there is a lot of material from space falling on the Earth. As large amounts fall as dust we don't notice the gradual build up of the planet although several thousand meteorites do fall to earth per year. The reason we don't notice the meteorites unlike the Chebarkul meteorite which was caught on camera, is that they generally fall in more isolated places far away from people.

When the Chebarkul meteorite was witnessed by people around the world the question was raised as to why some asteroids make a path towards the Earth. Almost all of the asteroids that we know of, which is over 95%, remain travelling on the main asteroid belt. The Asteroid Belt which is on orbit between Jupiter and Mars is over 200 million kilometres across. The asteroids within the asteroid belt have been following the same path for millions of years and as long as they stay in the asteroid belt following their same path they are no threat to the earth. It is only when an asteroid breaks from its set path through a collision or by light hitting the asteroid (Yakosky Effect, through analysis it is understood that particles of light can create a dramatic force over millions of years) which over time will eventually lead to the asteroid making a new path in which case can set the asteroid in a new direction toward the earth.

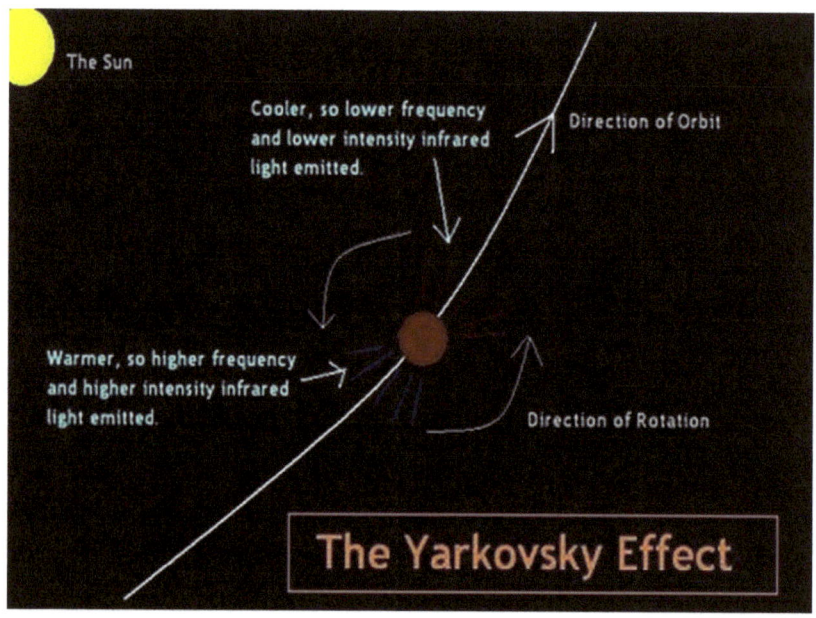

Astroprofs Page Oct 30th 2006

Jupiter is the largest planet in the Solar System and the fifth planet from the sun. It is classified as a gas giant and is 300 times bigger than the earth. It is the third brightest object in the night sky after Venus and the

Moon. Jupiter also has a huge gravitational field and has the largest planetary atmosphere in the Solar system. A very important aspect to Jupiter for the Earth is that it can draw or pull large object close to it and its atmosphere can tear apart asteroids with its gravitational pull. One other important aspect to Jupiter which is not so beneficial for the earth is that it can also deflect asteroids into orbit which crosses the Earth's path. Scientists believe that this is what caused Chebarkul meteorite. Initially it was thrown out of its orbit and may have spent millions of years travelling before it collided with the Earth.

Jupiter Mission Junocam NASA

The Chebarkul meteorite travelled at roughly 17.5 kilometres per second which is 65,000 kilometres per hour and measured 15 kilometres across. It entered the Earth's atmosphere at 9.20 am local time and arrived at 55degrees latitude north. The effects of the meteorite were felt over 3000 square kilometres. It exploded in an airburst at a height of 15 to 25 km and is the largest known object to enter the earth's atmosphere since 1908 when the Tunguska event occurred.

On June 30th 1908 a massive explosion occurred in Siberia and tore through the Forrest of Tunguska levelling over 16 million trees across an area the size of London. Through study and analysis it was discovered that the devastation was caused by an asteroid which was similar to the Chebarkul meteorite with the respect that it exploded over the area before it hit the ground. The explosion which created an air burst or blast wake with hurricane winds which tore down trees on over thousands of square kilometres. The explosion occurred near the Podkamennaya Tunguska River at 07.14am local time. It is the largest impact event on or near the earth in recorded history.

Trees felled by the Tunguska explosion. Credit: the Leonid Kulik Expedition.

When an asteroid enters the Earth's atmosphere at recorded speeds of up to 20 kilometres per second the Earth's atmosphere with its air resistance decelerates the asteroid and as it begins to slow down rapidly it begins to break apart causing a huge explosion. This explosion or blast wave with very high winds will bring as we saw with the Chebarkul meteorite a sonic boom that will cause destruction to the surrounding area and with the case of the Tunguska Event, a force with the strength of a powerful hurricane.

Scientists think Lake Cheko, in the Siberian region of Tunguska, may prove that a 1908 fireball was actually a meteor.

An eyewitness report from the Tunguska Event Testimony of S. Semenov, as recorded by Leonid Kulik's expedition in 1930:

At breakfast time I was sitting by the house at Vanavara Trading Post (65 kilometres/40 miles south of the explosion), facing north. I suddenly saw that directly to the north, over Onkoul's Tunguska Road, the sky split in two and fire appeared high and wide over the forest(as Semenov showed, about 50 degrees up-expedition note). The split in the sky grew larger, and the entire northern side was covered with fire. At that moment I became so hot that I couldn't bear it, as if my shirt was on fire; from the northern side, where the fire was, came strong heat. I wanted to tear off my shirt and throw it down, but then the sky shut closed, and a strong thump sounded, and I was thrown a few meters. I lost my senses for a moment, but then my wife ran out and led me to the house. After that such noise came, as if rocks were falling or cannons were firing, the earth shook, and when I was on the ground, I pressed my head down, feeling rocks would smash it. When the sky opened up, hot wind raced between the houses, like from cannons, which left traces in the ground like pathways, and it damaged some crops. Later we saw that many windows were shattered, and in the barn a part of the iron lock snapped.

The asteroid that levelled Tunguska could have been between the sizes of 30-100 meters in diameter which is quite small. We know from the events at Tunguska and Chelyabinsk that small asteroids can be very dangerous to the earth. From scientific analysis we know that there are a larger number of small asteroids than larger ones and we know now that not all of these asteroids can be detected in space while travelling toward the earth. The size and the speed of the object into the earth's atmosphere is what give's the asteroid its awesome power. If the asteroid is small and explodes high in the earth's atmosphere there will be a minimal effect to the earth from the shockwave but if it is a large asteroid and travels low in to the atmosphere the effects can be very destructive and the shockwave can be devastating.

Tunguska Event. The morning of June 30th 1908

What we consider to be very devastating to the earth and its environment such as large earthquakes and volcanic eruptions can seem trivial to the force than can be unleashed by a meteorite. The Barringer crater or Canyon Diablo crater in Arizona is the first confirmed asteroid crater on earth. From analysis we can tell that the blast from this meteorite would have devastated a very large area or city the size of London. The impact can be traced to 50,000 years ago. The size of the meteorite was roughly 50 meters across and left a crater 1 kilometre across and 200 meters deep. From the measurements of the crater and its depth we can tell that it was about the same size as the meteor that struck Tunguska, Siberia. The main difference between the two meteors is that this meteor struck the ground.

March 9, 2005 from NASA shows the Meteor Crater in Arizona.

Ground Strikes from meteorites are amongst the most destructive natural hazards that can befall the earth. Asteroids can release far greater destruction than what they leave as their footprint when they reach the ground. When the meteorite entered our atmosphere in Chelyabinsk and the fact that it was an airburst limited the consequences for the people living in the area. When a meteorite hits the ground the kinetic energy that is delivered into the ground creates a seismic shock like an earthquake.

There is evidence to suggest that the earth was hit by a devastating meteorite that changed the course of life on the planet. The impact dates back to millions of years ago. The Cenote (local Spanish word for sink hole) ring which is in the northern Yucatan Peninsula is a semi circular boundary of fractured and un-fractured rock. It is believed by scientists that the origin of the ring of sink holes was caused by a buried impact crater. From the air and from mapping of the area the sink holes are visible as a nearly perfect circular feature. The ring stretches on for 170 kilometres with a gravimetric and magnetic high lying at the centre of the ring. There are no combination of stress faults that could have produced an almost circular ring and what is mapped through scientific instruments is that the rock has been deformed revealing the boundaries of a colossal meteorite impact centre.

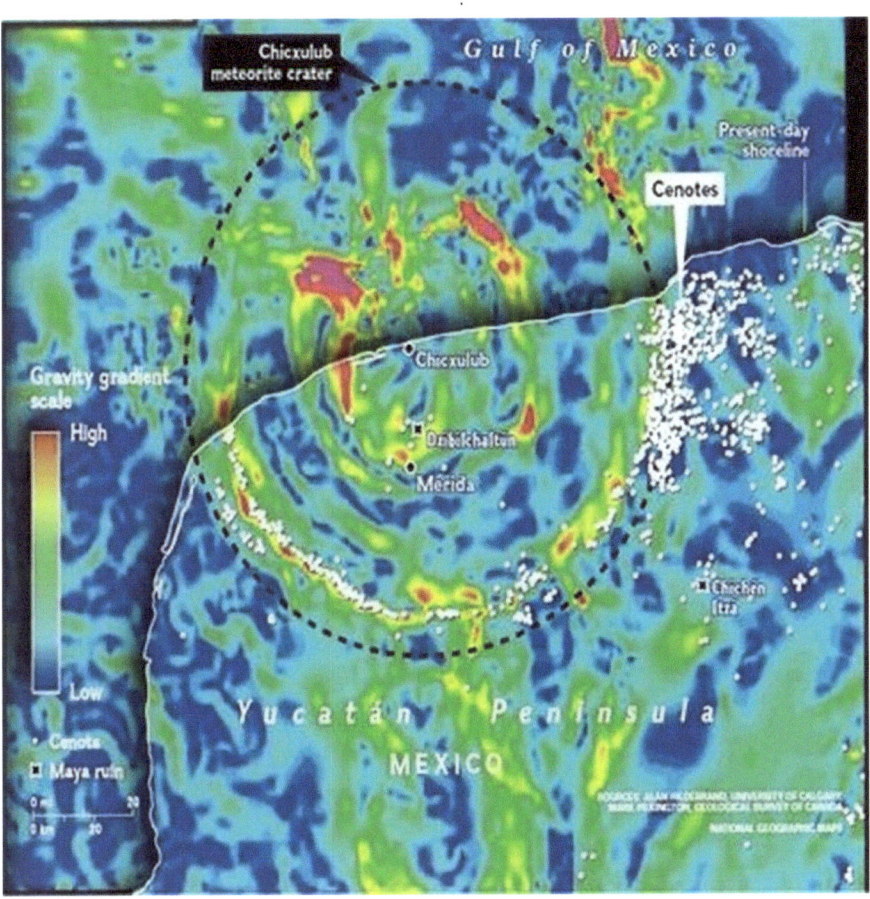

A meteorite slammed into the Yucatan Peninsula, causing a mass extinction event to begin.

The meteorite that struck the area was 15 kilometres across and it is believed that this meteorite affected the whole planet and led to the destruction of the dinosaurs. The impact is dated to 65 million years ago and caused utter destruction to the whole planet exploding with the force of 100 million tons of TNT. The force of the impact sent a giant plume of vaporised rock out into space which fell back to earth. There were billions of superheated molten particles falling back with a temperature of 100's of degrees these superheated the air causing massive fires that swept the planet. The atmosphere would have been laden with soot and dust which in turn would not allow for life to continue on the planet as it had previously done. Hence the dinosaurs and other life forms perished. The meteorite punched a crater 30 kilometres into the earth's crust.

New information from researchers in New Hampshire now believe that the space rock that fell on the Yucatan Peninsula in Mexico 65 million years ago was probably a speeding comet. Details in the 44th Lunar and Planetary Science conference outlined details to this belief but not all researchers are convinced. The iridium values on the impact area are said to be incorrect. Iridium is a chemical element, silvery white, brittle and hard. It is the most corrosion resistant metal and is a member of the platinum family. Iridium is found in meteorites consisting in much larger quantities than are found in the earth's crust. With a lower count of iridium at the impact site in the Yucatan Peninsula it left the researchers with the belief that it was not a meteorite that had collided with the Yucatan. Due to the collision causing a 180km wide crater it is understood that what had made the collision was travelling at a very high speed. The researchers believe that based on the evidence a long period comet is the most likely candidate for such a collision.

Halley's Comet

An event like this could strike the earth again but from mapping the asteroids in space we can see that the very biggest asteroids do not pose any immediate threat to the earth and they are few in number. The main worry for the earth which was demonstrated in Chelyabinsk is that there are a large number of small asteroids and lots of small rocks which could threaten the planet. The main nerve centre for asteroid detection is the Minor Planet Centre just outside of Boston. They have stressed in the past that every asteroid in the solar system has been accounted for but we know from the meteor strike above Chelyabinsk that this is not the case.

The function of the Minor Planet Centre is to follow the path of asteroids detecting any movement which will place the asteroid on a collision course with the earth. The nearest asteroids are constantly analysed to see if they are getting closer to the earth and there are over 9000 near earth asteroids accounted for. The sizes of the asteroids that are the major concern for the earth are the ones with a size of over 1kilmoetre in diameter. If an asteroid of this size was to hit the earth it would cause a major disaster for the planet. The Minor Planet Centre has detected over 900 asteroids that are over 1 kilometre in diameter but the good news for the planet is that they are maintaining their path on the asteroid belt and could not possibly collide with the planet for at least 100 years.

On the same day of the Chebarkul meteorite, 15th February 2013, another and much larger asteroid was about to pass the earth travelling between the earth and the communication satellites. Asteroid 2012 DA14 was similar in size to the meteorite that created the Barringer crater came within 28,000 kilometres of the earth. The asteroid passed inside the orbit of our chief stationary satellites before heading off to the north. It was the closest we have been on the earth to record an asteroid passing at this size. The asteroid had been successfully tracked for over a year and it was known that despite its proximity it would pose no threat to the earth.

Asteroid 2012 DA14 (streak at top center) is seen seven hours before its closest approach to Earth. Image courtesy Dave Herald, Murrumbateman, Australia

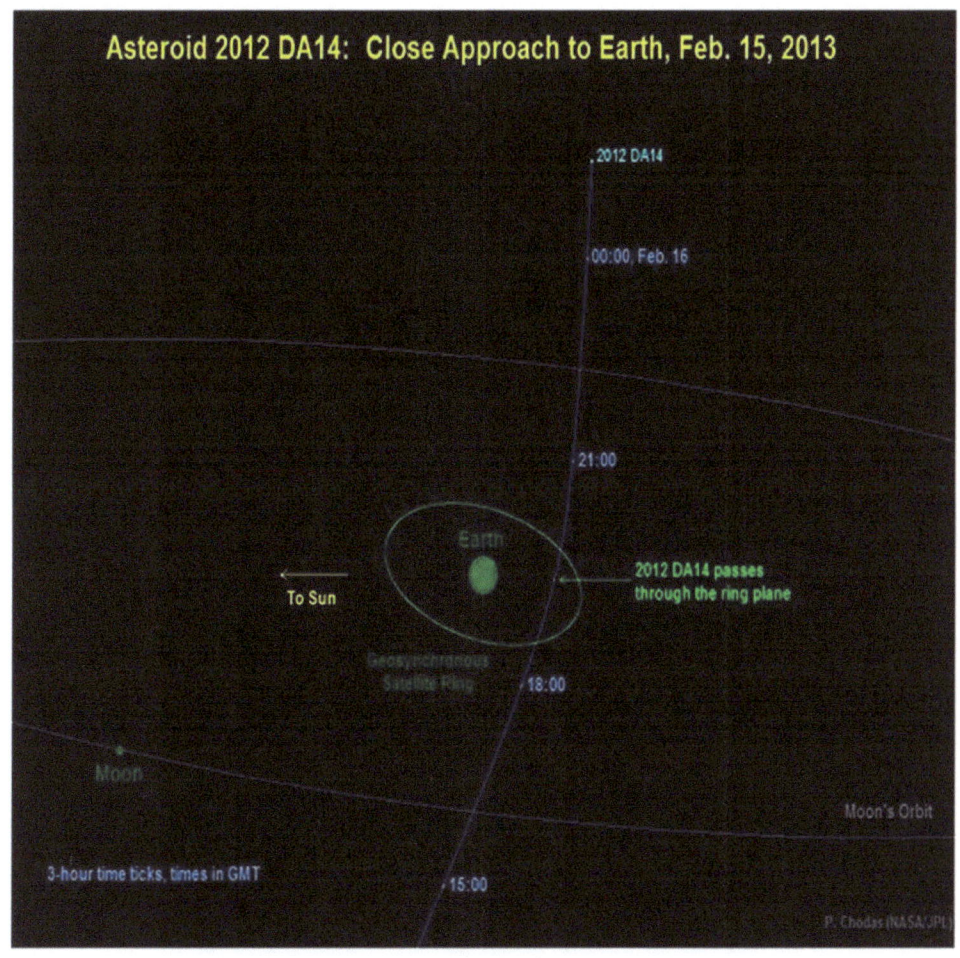

NASA/JPL Near-Earth Object Program Office

We can be reassured that there are no large asteroids over 1 kilometre that will impact the earth for at least 100 years but we are still faced with the threat of smaller asteroids like the one that entered our atmosphere over Chelyabinsk. We do not have a warning for the smaller asteroids as they are too hard to detect in space and there are some asteroids that we do not know where they are.

Comets

There are over 400,000 million stars in our galaxy and where the stars are most tightly paced together can cause problems for the earth. Where the sun goes the earth goes. As the sun travels through the galaxy we pass many stars. Where they are densely packed together is where the greatest danger for the earth lies. The gravity from the other stars on the solar system causes the earth problems. In outer space there is a cloud of chunks of ice consisting of trillions of lumps of ice moving within the cloud. If for example another star comes too close to the cloud of ice the stars gravity disturbs the cloud. This can then lead to ice chunks moving towards the sun. Some of these ice chunks can be huge in diameter and travel at speeds of 40 Km's per second.

On 30 April 2013, NASA's Hubble Space Telescope observed Comet ISON. This beautiful composite image was created from the images made by the HST. (Image Credit: NASA, ESA, and the Hubble Heritage Team (STScI/AURA)

Components of Comets

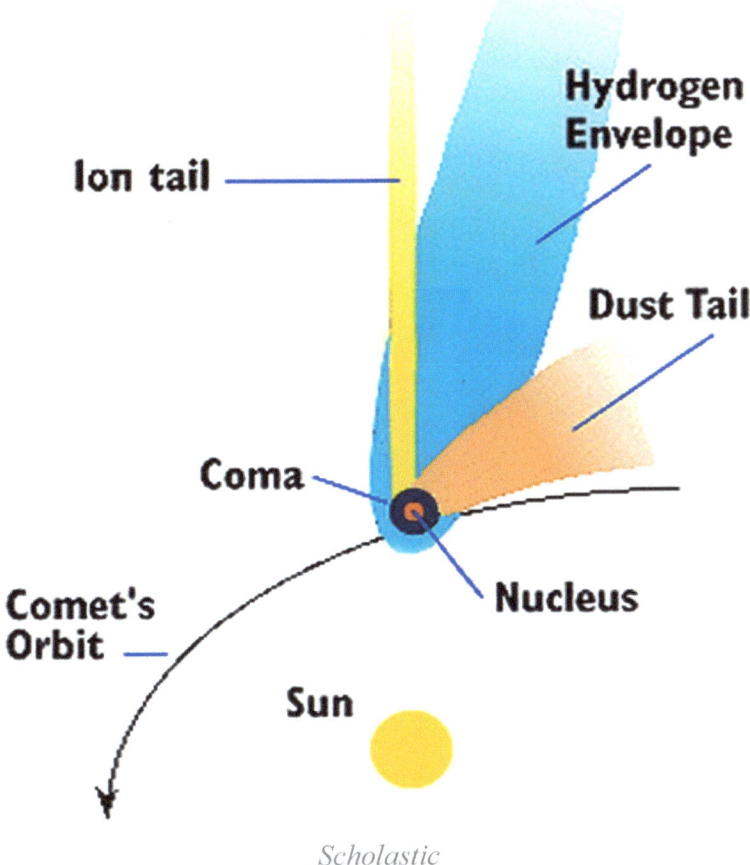

Scholastic

    The lump of ice or comet begins a journey through the heart of our solar system. These lumps of ice are called comets. As the comet moves steadily towards the sun it warms up and releases a haze of gas and dust. A cloud travels through space behind the comet and this is called the comets tail. The comet that hit the Yucatan Peninsula did not hit the sun but skimmed around it and passed through Mercury and the planets gravity directed the comet on a path with the moon. The comet then would have passed the moon and on to a collision course with the earth. When the comet hit the Yucatan Peninsula it would have caused enormous tsunamis and destabilised the earth's atmosphere. Whether what hit the earth at that time was indeed a comet or a meteorite the ultimate result was that most life on earth perished including the dinosaurs.

Space Station Space Comets

We must be aware that the earth has a history of impact with meteorites and comets over a period of billions of years and they are just part of the life cycle of the planet earth.

www.ingramcontent.com/pod-product-compliance
Lightning Source LLC
Chambersburg PA
CBHW040851180526
45159CB00001B/383